Block Courtesy] [*The Railway Gazette*

Diesel Parcels Car No. 34

(*see page* L20)

THE LOCOMOTIVES OF THE GREAT WESTERN RAILWAY

Part Eleven

The Rail Motor Vehicles and Internal Combustion Locomotives

First Edition - - - 1952
Second Edition - - 1956

Published by
THE RAILWAY CORRESPONDENCE AND TRAVEL SOCIETY
1956

THE LOCOMOTIVES OF THE GREAT WESTERN RAILWAY

Part Eleven

CONTENTS

THE RAIL MOTOR VEHICLES AND INTERNAL COMBUSTION LOCOMOTIVES

Combined locomotives and carriages for ht passenger work were first used in this ıntry in 1847-49 on the Eastern Counties d Bristol and Exeter Railways, four exams in all being built, but the practice did not d general favour until the early years of e present century when Mr. Dugald Drumond revived the idea in some steam railotors built for the joint use of the London & uth Western and London Brighton & South ast Railways. Two cars were constructed the service between Fratton and Havant d one of these was loaned to the Great estern Railway for trials early in 1903. sed on the results of these trials, Mr. G. J. urchward designed some cars for the W.R. and this Company soon became the ost extensive user of steam railmotors in e country. Several of the G.W.R.'s smaller ighbours in South Wales also adopted these hicles for their branch services and some of e designs bore a striking resemblance to the W. cars.

Steam railmotors were originally intended supplement the train service for ort distances on main lines (often to meet ectric street tramway competition) and to ork the entire service on many of the less portant branches and were quite successful hen confined to these duties, especially if the affic was light enough to run the cars witht trailers. Their main weakness, inherent most self-propelled vehicles, was their lack flexibility on account of their limited commodation and power; on several serces they developed so much traffic that at e busier hours they had to be replaced by ains hauled by ordinary engines. It was idently found more economical to work e engine on all the trips in these cases, as number of small engines were soon equipped ith manual control gear enabling them to e driven from the other end of trailer coaches milar to those used with the steam railotors. Up to four coaches could be handled pairs on either side of the engine and this ade it much easier to adjust the accomodation to the requirements of the traffic; e arrangement retained one of the railotor's advantages in that the engine did ot need to run round the train at the end of ach trip.

In the course of a speech at the August 903 half-yearly meeting, the G.W.R.'s hairman told the Directors that the alteratives of steam and petrol propulsion had een considered and that a decision had been ade in favour of steam. He announced that he cars would not only serve existing stations ut (with the sanction of the Board of Trade) would also pick up passengers at level crossings. This was evidently allowed as there were several examples of the cars being approached from rail level by means of special retractable steps. Most of these places were later provided with platforms of standard height, although the "halts" formed by one of the constituent companies between Pontypridd, Caerphilly and Machen in South Wales are mostly still at rail level.

The G.W.R. first experimented with internal combustion engined vehicles in 1911, when a small petrol-electric railcar was put into service but apart from a few petrol shunting engines and gangers' trolleys purchased in the 1920's, they did nothing further in this field until 1933, when the first diesel-mechanical railcar for passenger service was introduced. Thirty-seven more cars have since been added, including some with buffet facilities for express services and some built entirely for parcels traffic; the last twenty-one to be built are adapted for trailer haulage and multiple-unit working.

A small diesel-mechanical shunter was purchased for working in the yards at Swindon and a larger diesel-electric shunter bought for use in the Acton marshalling yard. Seven further diesel-electric shunting engines generally similar to the latter were put in hand by the G.W.R. but had not been completed when the Company was merged into British Railways. Similarly, two experimental Gas Turbine Locomotives ordered at the end of the War had not been completed by 31st December, 1947.

The Engineering Department makes quite extensive use of small 4ft. 8½in. gauge petrol-driven trolleys for permanent way maintenance work. These are arranged so that they can be lifted off the track at selected points where provision is made for berthing them clear of the ordinary traffic and they are also capable of drawing a small trailer. No further details of these vehicles are given as they hardly fall to be dealt with as locomotives.

From 1906 to 1938, the G.W.R. had a half share in the forty electric motor coaches of the Hammersmith and City Railway jointly with the Metropolitan Railway. These were numbered 2208-2247 when included in the London Passenger Transport Board stock and were the G.W.'s first and only venture into normal railway electrification.

Through the courtesy of the makers, it has been possible to include full details and photographs of most of the absorbed companies' railmotors, much of this information not having been published before; accounts published in the *Locomotive Magazine* when the various cars entered service have also provided useful information.

Many of the G.W. and absorbed companies' railmotors were converted to trailers ; a chapter is included on these and on the various types of trailer built to work with the railmotors ; the auto engines which also worked with these trailers are briefly described. Details of individual engines equipped for auto working will be given when dealing with the appropriate engine classes.

This section of the Great Western Railway's locomotive history has been compiled by Mr. F. K. Davies on behalf of the Society, mostly based upon official information kindly made available by the G.W.R. and the Western Region of British Railways. Thanks are also due to the following firms who have supplied details and photographs : Gloucester Railway Carriage and Wagon Co., Ltd. ; Hudswell, Clarke & Co., Ltd. ; The Hunslet Engine Co., Ltd. (incorporating Kerr, Stu & Co. and the Avonside Engine Co.) ; Mo Rail, Ltd. ; North British Locomotive C Ltd. ; Robert Stephenson & Hawthor Ltd. (incorporating Manning, Wardle & C Ltd. and Kitson & Co., Ltd.) ; W. Sis & Co., Ltd. ; The Associated Equipment C Ltd. ; George Cohen Sons & Co., Ltd. ; Le Bros., Ltd., Port Sunlight ; The Brit Thomson-Houston Co., Ltd. ; The Mauds Motor Co., Ltd. Members of the Railw Correspondence and Travel Society have a given invaluable help, especially Mr. E. Mountford, who has discovered a great d of official data on the Absorbed Compani railmotors which is published for the fi time in this new edition.

The information in this Part has be revised to include all known changes November, 1955.

THE G.W.R. STEAM RAIL MOTORS

The first two Great Western steam railmotors were built at Swindon and put into service between Stonehouse and Chalford on 12th October, 1903, and by 1908 no fewer than ninety-nine coach bodies and one hundred and twelve engine portions had been built, the balance of engines being used as spares. Engines Nos. 0864-0877 were built by Kerr, Stuart & Co., Makers' Nos. 906/7/31-42, Nos. 0853-8 at Wolverhampton and the remainder at Swindon.

The engine unit numbers 0801-0912 were part of a series allotted to various machines and had no connection with ordinary locomotive numbers. These numbers were engraved, together with the building date, on an oval brass plate fixed to the bogie stretcher between the two cylinders. Engines built in G.W.R. workshops had no Works Numbers but the following details are recorded :—

Order or Voucher Nos.		*Engine Nos.*
F.23	1 & 2	0801/2
F.37	1-12	0803-14
F.44	1-12	0815-26
231	1-24	0827-50
H.80/1		0851/2
231	25-30	0853-8
103	1-3	0859-61
114	1 & 2	0862/3
K.S. Carr. Lot	1099	0864/5
K.S. 158	1-12	0866-77
Gloucester 161	1-8	0878-85
H.109	1 & 2	0886/7
H.168	1-6	0888-93
Carriage Lot	1129	0894-6
Carriage Lot	1142	0897-912

The coach bodies were numbered separately ; Nos. 15, 16 and 61-72 were built by Kerr Stuart & Co. (with the engines given above), Nos. 73-80 by the Gloucester Carria & Wagon Co. (*for* engines 0878-0885, built Swindon) and the remainder at Swindo under the following carriage Lot Numbers :

Lot Number	*Coach Numbers*	*Lot Number*	*Coach Numbers*
1037	1, 2	1099	15, 16
1054	3-14	1100	61-72
1063	17-28	1101	73-80
1078	29-40	1129	81-83
1079	41-52	1140	84-90
1088	53-58	1142	91-99
1089	59, 60		

The building dates of the engines diff slightly from the dates of completion of t coach bodies into which they were fitted ar Table 1 gives details of dates and also show the *original* combination of engines an coaches. The engines and coaches interchange repeatedly, often at every repair, and it not practicable to give details of such change

Cars 15 and 16 (Engines 0864/5) differe considerably from all the others and wer probably designed by the makers. The leadin dimensions of these two engines were :—

Cylinders (2 outside) . .	9in. × 15in.
Driving wheels	3ft. 5in.
Carrying wheels	2ft. 10in.
Boiler, Barrel length . .	6ft. 1¼in.
,, Diameter . .	3ft. 8in.
Tubes, Number	172
,, Diameter . .	1½in.
Heating Surface, Tubes	307.75 sq. ft.
,, ,, Firebox	42.75 ,, ,,
Grate Area	7.25 ,, ,,
Working Pressure . .	170 lb. sq. in.
Water Capacity	338 gallons.
Coal Capacity	15 cwt.
Wheelbase, Engine Bogie	9ft. 9in.
,, Carrying ,,	8ft. 0in.

Photograph] [*British Railways*

G.W.R. Steam Rail Motor No. 1

L2

G.W.R. Steam Rail Motor No. 2
At one of the Rail level Halts
between Chalford and Stonehouse

L3

Photograph] [*Locomotive Publishing Co.*

Nidd Valley Light Railway Rail Motor *Hill*
at Lofthouse, Nidd Valley
Formerly G.W.R. Car No. 15 (0864)

L4

THE G.W.R. STEAM RAIL MOTORS
(*see page* L4)

Photograph] G.W.R. Steam Rail Motor No. 17 [*E. Pouteau*
At Brentford

L5

Photograph] [*H. C. Casserley*
Driving End of G.W.R. Steam Rail Motor No. 40
At Stourbridge, April, 1932

L6

Photograph] G.W.R. Steam Rail Motor No. 77 [*S. W. Baker*
At Swindon

THE G.W.R. STEAM RAIL MOTORS

L7

(*see page* L4)

Wheelbase, Total	..	50ft. 0½in.
Weight, full	.. ..	32 tons 0 cwt.
Length over mouldings	..	56ft. 3in.
Length over buffers	..	59ft. 9in.
Seating capacity	..	48

These details were given on Diagram E except the boiler length, which information was supplied by the makers. The illustration of ex-Car No. 15 at work on the Nidd Valley Railway shows that there was a transverse boiler and that the engine bogie wheels were not coupled ; the weight distribution on the Diagram therefore appears as 13 tons 12 cwt. + 9 tons 1 cwt., + 9 tons 7 cwt. The carrying bogie wheels had inside bearings.

The engine portions of all the other railcars had four-coupled wheels and two 12in. by 16in. outside cylinders with Walschaerts valve-gear operating overhead " D " slide valves. The wheels of Nos. 0801/2 were 3ft. 8in., of 0803-36 3ft. 6½in., and of all the rest 4ft. 0in. diameter. The coned vertical boilers were all 9ft. 6in. high with a minimum diameter of 4ft. 6in. the maximum diameter being 6ft. 0in. for Nos. 0801-0805 and 6ft. 0⅜in. for the remainder. The boilers of Nos. 0801/2 are shown on Diagrams A and A[1] as working at 140 lb. per square inch at first, later increased to 160 lb. and the figure of 180 lb. given in other accounts is not substantiated by official data. All later boilers worked at 160 lb. per square inch and the remaining boiler dimensions were as under :—

	0801/2	* 0803-5	0806-29	0830-50	0851-63 0866-96	0897-0912	§ Later.
Tubes, Number ..	477	333	477	462	420	419	317
,, Diameter ..	1⅛in.	1⅛in.	1⅛in.	1⅛in.	1¼in.	1¼in.	1⅝in.
Heating Surface, Tubes (sq. ft.) ..	625.58	436.72	625.58	605.90	612.78	611.32	601.22
Heating Surface, Firebox (sq. ft.)	44.34	38.82	46.75	46.75	46.46	46.46	46.14
Grate Area (sq. ft.)	11.48	8.4	11.54	11.54	11.54	11.54	11.54

*These engines appear to have been altered to conform to Nos. 0806-29 ; the original heating surface was probably found too small.

§ Several boilers were later retubed thus but as the Diagrams D[1], G[1] and J[1] showing these differences relate to the coaches, it is not practicable to say which engines were so treated.

The water capacity was 450 gallons in all cases and the engine bogie wheelbase was 8ft. 0in. The other dimensions varied with the cars into which the engine units were fitted and these are set out in Table 2. The actual engine unit numbers corresponding to each Diagram cannot be given with any degree of certainty, because of the frequent changes already mentioned. The weights indicated by small metal figures at the ends of the cars have been recorded by observation in some cases and these were generally a couple of tons higher than those shown on the Diagrams.

The steam railmotors were recorded on lettered diagrams in a similar manner to the locomotive stock, except that minor differences between cars nominally identical were shown on separate diagrams with the same index letter. For example, Car No. 1 shown on Diagram A had a partition forming a smoking compartment, while Car No. 2 (Diagram A[1]) had no internal divisions. The complete list of diagrams is included and it should be noted that they are listed under *coach numbers*. The changes of engine portions were only recorded by alterations to Diagram Letters in a few cases and it has been found convenient to tabulate most of the dimensions separately from the Diagram list.

In 1905, long before the last coaches were built, the G.W.R. started fitting small tank engines (mostly " 517 " class 0-4-2T) with auto gear and many of the ninety-seven trailers intended for pairing with the steam cars were used in conjunction with these engines instead. The use of separate engines and trailers was found more satisfactory (two pairs of trailers could be handled by an engine when the traffic demanded, whereas the cars sometimes had difficulty in taking one trailer and keeping time) and the first railmotors were withdrawn in 1914. They disappeared gradually during the next twenty-one years and the last service to be operated by them was between Neath (Canal Side) and Court Sart. This ceased on 16th September, 1935, and that year saw steam railmotors removed completely from G.W. stock. Coach No. 45 was on loan to the Highland Railway from January, 1918, to March, 1924, but it is not known which engine unit was in it at the time.

Engines and coaches were withdrawn separately, with the exception of three which were sold as complete units, and the number of spare engines varied from year to year. The annual stock returns showed the number of existing coaches as the railcar total and any excess of engines was shown as " spares." Table 3 shows the date when each engine was withdrawn from service and also the number of the coach in which the engine last worked. This does not necessarily mean that it worked in that coach until the date given, as it may have been a " spare " for a period immediately prior to withdrawal.

With the exception of the three sold as complete units, one destroyed by fire and eleven broken up, the coach bodies were converted into trailer cars and in Table 4 the resulting trailer car numbers are given where applicable. It will be noticed that the last ten cars to be withdrawn were all broken up.

The whole of the foregoing dates, dimensions, etc., and Tables 1, 2, 3 and 4 have been reproduced from Mr. N. J. Allcock's manuscript, having been compiled by him from official information at Swindon through the courtesy of Messrs. C. B. Collett and F. W. Hawksworth, the Chief Mechanical Engineers concerned. This complete account supersedes the details given on pages 131/2 of the *Railway Observer* for May, 1937, and in the *Railway Magazine* for June, 1929. In particular, the statement in both these articles that Cars Nos. 1 and 2 originally worked at 180 lb. pressure is not correct.

In connection with the cars sold to the Port of London Authority, it is understood that Nos. 1 and 2 were overhauled after purchase and received some alterations, the gas cylinders being removed and a dynamo for generating electric light fitted. Their weight was also reduced and other adjustments made to suit them for work on the Millwall Extension Railway. This line was closed for passengers in 1926 and the railmotors were said to have been sold to Messrs. T. W. Ward in November, 1927, but they were advertised for sale on 19/8/28, being then on the Glenfield Siding. They were broken up there, the last remnants being removed in late 1928. It is possible, of course, that Ward's bought them but were unable to dispose of them and broke them up on the spot.

It is interesting to note that Car No. 43 (G.W. built "Suburban" type) was illustrated in Messrs. Kerr, Stuart's catalogue, although they actually built Cars 61-72, all of which were of the "Branch" type.

There were considerable variations in detail between the cars but they fall into four main groups, two of which comprised only two cars each.

Nos. 15 and 16 were evidently of the maker's own design and their general appearance can be seen from the illustration of the Nidd Valley car (ex-G.W. No. 15) (fig. L4).

The first cars, Nos. 1 and 2, were also different from all the others of G.W.R. design; they were entered only from the leading end, via the driving compartment, by means of a fixed flight of steps to suit rail-level access and this vestibule had collapsible iron gates in place of the proper doors provided on later examples. These two cars had flat ends but the later cars had bow ends.

The sides of Nos. 1 and 2 were of vertical matchboarding, a feature shared by Nos. 3-14 and 17-28. A photograph of one of this series (fig. L5) is reproduced and it will be seen that this has steps to suit the usual platform heig only. Nos. 29-99 all had panelled sides, b ends and retractable steps to suit rail-le access. Nos. 59 and 60 had vestibuled conn tions at the leading end, to pair with traile similarly fitted.

The G.W.R. cars were all gas lit and no was ever altered to electric lighting, althou trailer conversions Nos. 126 (1929), 127, 14 181-5 (1930) and 134/5 (1931) were so fitt at the dates shown. They were all Third Cla only, although some at least were divided in Smoking and Non-Smoking compartments.

The arrangement of the passenger e trances varied, some cars being approache only through the driver's vestibule, othe (including Nos. 81-99) having single centr doors in addition, while Nos. 61-80 an perhaps others had the end entrance and pair of centre doors. The doors opened int vestibules which contained the levers fo working the retractable steps.

The difference between the "Suburban" and "Branch" cars was that the former ha no separate luggage compartment.

Coal was carried in a bunker in the bow end of the boiler compartment, while the water tanks, placed under the coach body were filled by means of pipes which terminated behind the oval plates on the sides of the boiler room, as shown in the photograph of No. 77. The bunkers on Cars Nos. 1 and 2 held only 10 cwt., those of Nos. 3 to 83 inclusive 15 cwt., and Nos. 84-99 30 cwt. The vacuum apparatus included a small ejector as no vacuum pumps were fitted.

Mr. A. E. Harris of Bristol was actively associated with the steam railmotors at St. Philip's Marsh shed and he writes about performance as follows :—" . . . when the car hauled a trailer and sometimes a van as well, the fireman was driven almost to distraction in his efforts to maintain steam, as he dare not open the firehole doors between stations with the regulator also open. In fact, the opposite to locomotive practice was observed, the cars had to be fired when the regulator was closed and if no favourable gradient was encountered, firing had to be done at a station stop, despite Rule 126 (V), to avoid a further pressure drop. Steam would then have to be used to inject water into the boiler, which by this time would probably have an uncomfortably low water level. Experienced firemen who had "passed out" in passenger train working were rostered for the rail motor turns as they had to be well acquainted with the drivers' duties; when the driver was at the other end of the coach, the fireman had to notch up the gear and create the vacuum. The driver, fireman and guard were in touch through a code rung on an electric bell."

" In their early days the cars were painted crimson lake, wheels included, although when the chocolate and cream livery was adopted for the coach portion, the wheels (like the rest of the undercarriage) were painted black. The only modification I ever saw on the rail-cars was the provision of a large warning gong at the leading end, operated by a foot pedal as the whistle at the boiler end was not fully effective when the engine was at the rear . . . I believe the cars were fairly light on repairs in relation to mileage and the two chief weaknesses were the loss of steam through strained steam delivery pipes around the socket joint or flange by the steam chest, and the fireman's dread—leaky tubes . . . When running solo, a railmotor would develop a good turn of speed but with a trailer the angry protests of the car could be felt by a constant jerking, especially at normal cruising speed. Railmotors always carried the " B " headlamp code (denoting stopping trains), this lamp being supplied by the Loco. Department while the tail lamp was always provided by the Carriage Department ! "

From the foregoing notes, it will be seen that the steam rail motors were successful when confined to the work for which they were first designed, that is for solo work on branch lines in quiet areas. Like the diesel railcars later described, they frequently developed traffic beyond their own capacity and in a way their success was their undoing. While quite a number lasted until the late 1920's, it was found desirable to replace them gradually by auto engines and some also became redundant because road motor competition forced the closing of the branches on which they worked.

TABLE 1

Coach No.	*Completion Date*	*First Engine*	*Date Built*	*Coach No.*	*Completion Date*	*First Engine*	*Date Built*
1	10/03	0801	10/03	41	1/05	0836	1/05
2	,,	0802	,,	42	,,	0834	,,
3	4/04	0803	3/04	43	2/05	0837	,,
4	,,	0804	4/04	44	,,	0839	,,
5	,,	0805	,,	45	,,	0840	2/05
6	,,	0806	,,	46	,,	0842	,,
7	5/04	0809	5/04	47	,,	0843	,,
8	,,	0810	,,	48	3/05	0845	,,
9	,,	0811	6/04	49	2/05	0841	,,
10	,,	0812	,,	50	3/05	0846	,,
11	6/04	0814	,,	51	4/05	0850	3/05
12	,,	0815	,,	52	2/05	0838	1/05
13	,,	0816	,,	53	9/05	0859	7/05
14	7/04	0818	7/04	54	,,	0860	8/05
15	10/05	0864	8/05	55	10/05	0855	4/05
16	11/05	0865	,,	56	,,	0856	5/05
17	4/04	0807	4/04	57	9/05	0861	8/05
18	,,	0808	,,	58	,,	0857	5/05
19	7/04	0820	8/04	59	11/05	0862	8/05
20	8/04	0824	,,	60	10/05	0853	3/05
21	6/04	0817	7/04	61	3/06	0866	1/06
22	7/04	0822	8/04	62	4/06	0867	,,
23	8/04	0823	,,	63	,,	0868	2/06
24	7/04	0821	,,	64	,,	0869	,,
25	5/04	0813	6/04	65	,,	0870	,,
26	6/04	0819	7/04	66	5/06	0871	,,
27	8/04	0825	8/04	67	6/06	0872	3/06
28	9/04	0826	,,	68	,,	0873	5/06
29	1/05	0835	1/05	69	,,	0874	,,
30	,,	0833	,,	70	,,	0875	,,
31	12/04	0827	11/04	71	,,	0876	,,
32	,,	0828	,,	72	7/06	0877	6/06
33	,,	0830	12/04	73	4/06	0878	2/06
34	1/05	0832	,,	74	,,	0879	,,
35	12/04	0829	,,	75	5/06	0880	4/06
36	,,	0831	,,	76	4/06	0881	,,
37	2/05	0844	2/05	77	,,	0882	5/06
38	3/05	0848	3/05	78	6/06	0883	5/06
39	4/05	0849	,,	79	,,	0884	6/06
40	3/05	0847	,,	80	,,	0885	,,

TABLE 2. STEAM RAIL MOTOR DIAGRAMS

Diagram	Coach Numbers	Classification	Carrying Wheel diameter (f)	Carrying Bogie Wheelbase	Total Wheelbase	Weight on Engine Bogie T	C	Weight on Carrying Bogie T	C	Total Weight full T	C	Length over mouldings	Length over buffers	Seating Capacity
A	1	Suburban	3′ 7″	8′ 6″	45′ 6″	24	0	12	0	36	0	57′ 0¾″	60′ 0¾″	52
A^{1}	2	,,	,,	,,	,,	,,		,,		,,		,,	,,	,,
B	3-5 (a)	,,	,,	,,	48′ 9″	22	13	11	12	34	5	59′ 6¾″	63′ 6¾″	,,
C	6-8	,,	,,	,,	,,	24	16	11	10	36	6	,,	,,	,,
D	9-14	,,	,,	,,	,,	,,		,,		,,		59′ 6″	,,	54
D^{1}	10, 14	,,	All dimensions as D, heating surface varied.											
F	17, 18, 25	Branch	3′ 7″	8′ 6″	48′ 9″	24	16	11	10	36	6	59′ 6″	63′ 6″	51
G	19-24, 26, 28	,,	,,	,,	45′ 9″	,,		,,		,,		,,	,,	,,
G^{1}	27	,,	3′ 7½″	,,	,,	,,		,,		,,		,,	,,	,,
H	29, 30/3/4/6	,,	,,	8′ 0″	46′ 6″	25	4	13	12	38	16	,,	,,	49
J	31, 35 (b)	,,	All dimensions as H, heating surface varied.											
J^{1}	32 (b)	,,	3′ 7½″	8′ 0″	46′ 6″	21	6	12	14	34	0	59′ 6″	63′ 6″	49
K	38-40	,,	,,	9′ 0″	57′ 2″	26	18	16	5	43	3	70′ 0″	74′ 0″	63
K^{1}	37 (c)	,,	,,	,,	,,	,,		,,		,,		,,	,,	,,
L	41, 42	Suburban	,,	8′ 0″	46′ 6″	25	4	13	12	38	16	59′ 6″	63′ 6″	50
M	44-48	,,	,,	9′ 0″	57′ 2″	26	18	16	5	43	3	70′ 0″	74′ 0″	64
M^{1}	43 (e)	,,	,,	,,	,,	,,		,,		,,		,,	,,	,,
N	49-52	,,	,,	,,	,,	,,		,,		,,		,,	,,	,,
O	53-8, 61-72	Branch	,,	,,	59′ 5″	,,		,,		,,		,,	,,	61
P	59, 60 (d)	,,	,,	,,	57′ 2″	,,		,,		,,		,,	,,	63
Q	73-80	,,	,,	8′ 0″	48′ 11″	25	4	13	12	38	16	59′ 6″	63′ 6″	45
Q^{1}	81-3	,,	,,	,,	,,	,,		,,		,,		,,	,,	,,
R	84-99	,,	,,	9′ 0″	59′ 5″	29	3	16	8	45	11	70′ 0″	74′ 0″	61
T	59, 60	,,	All dimensions as P, heating surface varied.											

NOTES : (a) Later Diagram C.
(b) ,, ,, H.
(c) ,, ,, K.
(d) ,, ,, T.
(e) Diagram M^{1} differed from M only in the arrangement of the wi dows.
(f) All carrying wheels had outside bearings.
For Diagram E, see pages L4 and L5.

Photograph] [*British Railways*

Bristol and Exeter Railway Steam Rail Motor *Fairfield* L8

Photograph] [*D. Clayton's Collection*

Alexandra Docks and Railway Steam Rail Motor No. 1 L9

Photograph] [*D. G. Coakham's Collection*

Alexandra Docks and Railway Steam Rail Motor No. 2 L10

THE STEAM RAIL MOTORS OF THE CONSTITUENT COMPANIES

(*see page* L12)

Photograph] [*North British Locomotive Co., Ltd.*

Barry Railway Steam Rail Motor No. 2

L11

Photograph] [*Gloucester Railway Carriage and Wagon Co., Ltd.*

Cardiff Railway Steam Rail Motor No. 3

L12

Photograph] [*R. Stephenson and Hawthorns, Ltd.*

Port Talbot Railway Steam Rail Motor No. 1

L13

THE STEAM RAIL MOTORS OF THE CONSTITUENT COMPANIES

(*see pages* L12 *and* L13)

Coach No.	*Completion Date*	*First Engine*	*Date Built*
81	5/07	0890	4/07
82	,,	0889	3/07
83	,,	0876*	5/06
84	12/07	0899	11/07
85	,,	0900	,,
86	,,	0897	10/07
87	,,	0898	,,
88	1/08	0902	11/07
89	12/07	0903	,,
90	,,	0901	,,
91	1/08	0905	,,
92	,,	0904	,,
93	2/08	0906	2/08
94	,,	0907	,,
95	,,	0908	,,
96	,,	0910	,,
97	,,	0912	,,
98	,,	0909	,,
99	,,	0911	,,

Coach No.	*Completion Date*	*First Engine*	*Date Built*
†		0851	11/04
†		0852	,,
†		0854	7/05
†		0858	10/05
†		0863	11/05
†		0886	4/06
†		0887	8/06
†		0888	3/07
†		0891	,,
†		0892	5/07
†		0893	9/07
†		0894	5/07
†		0895	6/07
†		0896	10/07

*This engine had previously been in Coach No. 71, which presumably now received one of the " spares."

†These fourteen engines were " spares " when built.

TABLE 3

Engine	*W'dn.*	*Coach*
0801	11/14	1
0802	,,	2
0803	9/22	20
0804	10/20	21
0805	9/19	24
0806	,,	28
0807	7/20	42(*a*)
0808	8/16	18
0809	,,	28
0810	,,	92
0811	10/19	23
0812	9/22	41
0813	9/19	29
0814	8/16	35
0815	10/20	18
0816	10/19	24
0817	10/20	31
0818	8/14	12
0819	8/16	19
0820	9/28	30
0821	9/22	34
0822	8/30	41
0823	10/19	25
0824	9/22	41
0825	10/22	42
0826	9/22	41
0827	4/28	41
0828	9/19	25
0829	,,	30
0830	9/22	13
0831	10/20	35
0832	,,	36
0833	9/22	14
0834	8/16	9
0835	9/19	27
0836	,,	32
0837	5/35	88
0838	10/33	96
0839	10/35	92
0840	5/34	71
0841	6/33	82
0842	10/35	65
0843	6/33	86
0844	12/34	72
0845	6/22	70
0846	6/35	71
0847	6/33	73
0848	12/34	80
0849	1/30	78
0850	12/29	75
0851	9/22	19
0852	9/19	22
0853	2/28	45
0854	5/28	69
0855	10/28	58
0856	12/34	81
0857	8/17	88
0858	7/20	49(*b*)
0859	9/22	68
0860	12/34	66
0861	6/33	58
0862	12/29	62
0863	12/34	64
0864	4/20	15(*c*)
0865	12/27	16(*d*)
0866	2/26	76
0867	6/14	85
0868	2/28	78
0869	12/34	96
0870	6/28	62
0871	5/28	65
0872	4/35	97
0873	12/34	93
0874	12/26	92
0875	11/14	62
0876	4/28	71
0877	1/35	76
0878	12/26	91
0879	6/28	53
0880	9/22	87

Engine	W'dn.	Coach
0881	7/14	45
0882	1/35	91
0883	1/30	81
0884	6/33	40
0885	,,	69
0886	9/22	33
0887	9/19	26
0888	12/34	77
0889	5/34	75
0890	6/35	37
0891	8/30	30
0892	4/35	30
0893	9/22	17
0894	8/28	39
0895	5/28	99
0896	1/35	70
0897	5/34	79
0898	10/29	94
0899	10/35	88
0900	6/33	74
0901	,,	53
0902	2/30	81
0903	5/28	87
0904	1/30	93
0905	2/30	64
0906	10/33	86
0907	5/35	55
0908	,,	98
0909	6/33	39
0910	12/26	54
0911	4/28	40
0912	6/33	83

(*a*) Sold as complete unit to Port of Lond Authority, becoming P.L.A.1, and scrapp 1928.

(*b*) Sold as complete unit to Port of Lond Authority, becoming P.L.A.2, and scrapp 1928.

(*c*) Sold as complete unit to J. F. Wake, 5/2 later resold to Nidd Valley (Bradfo Corporation) Waterworks and named *H* Sold for scrap June, 1937, and broken 10/37.

(*d*) Was to have been sold as complete ur in 1920, but was reinstated in stock 1/2

Note.—The two engines withdrawn from se vice in 12/29 and Nos. 0844/48/60/63/88 12/34 were not taken out of stock un years ending 12/30 and 12/35 respectivel

TABLE 4

Coach	W'dn.	Trailer
1	6/17	105
2	1/17	106
3	6/15	99
4	1/15	100
5	5/15	101
6	1/15	102
7	,,	103
8	2/15	104
9	2/16	107
10	12/16	108
11	3/17	109
12	12/16	110
13	3/19	111
14	11/18	112
15	4/20	(*e*)
16	12/27	(*f*)
17	4/19	113
18	5/19	114
19	8/19	115
20	9/19	116
21	5/20	117
22	2/20	118
23	3/20	119
24	5/20	120
25	4/20	121
26	2/20	122
27	3/20	123
28	2/20	124
29	3/20	125
30	10/35	—
31	12/22	129
32	4/20	128
33	12/22	130
34	,,	131
35	,,	132
36	1/23	133
37	10/35	—
38	12/27	146
39	6/33	197
40	,,	198
41	12/27	147
42	7/20	(*g*)
43	4/23	134
44	,,	135
45	12/27	148
46	11/22	136
47	10/22	137
48	1/16	(*h*)
49	7/20	(*k*)
50	4/23	138
51	,,	139
52	,,	140
53	6/33	199
54	1/30	181
55	10/35	—
56	1/30	182
57	12/27	149
58	6/33	200
59	6/20	126
60	,,	127
61	12/27	150
62	11/29	186
63	12/27	151
64	1/35	216
65	10/35	—
66	3/35	217
67	12/27	152
68	,,	153
69	6/33	201
70	10/35	—
71	10/35	—
72	1/35	218
73	6/33	202
74	,,	203
75	6/34	207
76	1/35	219
77	10/35	—
78	6/34	208
79	,,	209
80	10/35	—
81	11/34	211
82	6/33	204
83	,,	205
84	3/30	183
85	12/27	154
86	6/33	206
87	12/27	155
88	10/35	—
89	12/27	156
90	,,	157
91	1/35	210
92	10/35	—
93	11/34	212
94	4/30	185
95	3/30	184
96	1/34	213
97	4/35	214
98	6/35	215
99	12/27	158

(*e*) Sold with engine 0864. (*f*) Was to have been sol with engine 0865. (*g*) Sol with engine 0807. (*h*) Destroyed by fire. (*k*) Sold wit engine 0858.

Note.—The Rail Motor coaches withdrawn from service in 12/22, 12/27 and 11/29 were not withdrawn from stock until years ending 12/23, 12/28 and 12/30 respectively.

The steam railmotors were distributed over the greater part of the G.W. system and there were comparatively few branch lines over which they did not work. In some cases they were responsible for the entire service while in others they were used indiscriminately with auto trains, a practice which increased in later years when their numbers were dwindling.

Quite a number of successful suburban services on the main lines were initiated by the railmotors, the original Stroud Valley run (since extended to Gloucester) still being well patronised in spite of road competition.

Table 5 gives a representative selection of the services worked by the cars, with the sheds by which they were maintained or from which they operated ; it has not been possible to give a full list as the public timetables of the period do not differentiate between railmotors and auto trains. The table is based upon a list kindly furnished by ex-Inspector G. H. Partridge, M.B.E., who recently retired from the G.W.R., supplemented by the recollections of several members of the Railway Correspondence and Travel Society.

TABLE 5

SERVICES WORKED BY G.W.R. STEAM RAILMOTORS

Shed	*From*	*To*
LONDON DIVISION		
Southall	Southall	Brentford
	West Drayton	Uxbridge (Vine St.)
	,, ,,	Staines
	Gerrards Cross	Uxbridge (High St.)
	Westbourne Park	Ruislip & Ickenham (*via* Perivale and *via* West Ealing)
	,, ,,	Southall (*via* Perivale)
	Slough	Windsor
	Park Royal	Willesden Junction
Reading	Reading	Henley-on-Thames
	,,	Maidenhead
	,,	Didcot
	Bourne End	Marlow
Newbury	Newbury	Lambourn
Oxford	Oxford	Princes Risborough
BRISTOL DIVISION		
St. Philips Marsh	Bristol (Stapleton Road)	Clifton Down
	Bristol (Temple Meads)	Avonmouth
	,,	Portishead
Westbury and Frome	Westbury	Castle Cary
	Castle Cary	Taunton
	Westbury	Chippenham
	,,	Patney & Chirton
	,,	Warminster
	Trowbridge	Devizes
?	Chippenham	Calne
Bristol & Bath	Bristol	Chippenham
Bath	Limpley Stoke	Hallatrow
Yatton	Yatton	Swindon
	,,	Clevedon
	,,	Highbridge
	,,	Wells
Weymouth	Weymouth	Dorchester
	,,	Abbotsbury
NEWTON ABBOT DIVISION		
Exeter	Exeter	Dulverton
	,,	Dawlish Warren
	,,	Heathfield
Laira and Millbay	Plymouth	Saltash
	,,	Yealmpton
	,,	Tavistock
NEWTON ABBOT DIVISION—*continued.*		
St. Blazey	Lostwithiel	Fowey
Taunton	Taunton	Milverton
	Taunton	Castle Cary
Truro	Truro	Newquay *via* Perranporth
WOLVERHAMPTON DIVISION		
Banbury	Banbury	Princes Risborough
	,,	Chipping Norton
Birkenhead	Rock Ferry	Ledsham
Croes Newydd	Wrexham	Rhos
	,,	Coedpoeth
	,,	Moss
	,,	Llangollen
Much Wenlock	Much Wenlock	Craven Arms
Stourbridge	Stourbridge Junction	Stourbridge Town
	,,	Wolverhampton *via* Himley
	Dudley	Old Hill
?	Langley Green	Oldbury
	Old Hill	Halesowen
Tyseley	Tyseley	North Warwickshire Line
WORCESTER DIVISION		
Chalford	Chalford	Stonehouse
Cheltenham	Cheltenham	Honeybourne
Gloucester	Newnham	Forest of Dean
Kidderminster	Kidderminster	Bewdley
Honeybourne	Honeybourne	Stratford-on-Avon
	,,	Evesham
NEWPORT DIVISION		
Pontypool Road	Pontypool Road	Monmouth
	,,	Oakdale
Aberdare	Aberdare (Low Level)	Dare Valley Branch
	Black Lion Crossing	Cwmaman
Merthyr (?)	Merthyr	Newport *via* Nine Mile Point
NEATH DIVISION		
Neath	Neath (Canal Side)	Court Sart
	Glyn Neath	Swansea East Dock
Tenby	Pembroke Dock	Tenby
Llanelly (?)	Garnant	Gwaun-cae-Gurwen
Goodwick	Fishguard Harbour	Clarbeston Road

THE STEAM RAILMOTORS OF THE CONSTITUENT COMPANIES

BRISTOL & EXETER RAILWAY (1848).

The first steam railmotor car to work over what later became part of the G.W.R. system was Bristol and Exeter No. 29, *Fairfield*, which was actually the second combined engine and carriage to be built in this country. It will be seen from the illustration that this was a very quaint affair with a vertical boiler surmounted by a very tall chimney, mounted on an open platform apparently articulated with the coach portion; although as the illustration appears to be an artist's impression, perhaps it is not too accurate. It was built for the Broad Gauge.

It was designed and constructed in 1848 by W. B. Adams at the Fairfield Works, Bow, to the order of Mr. C. H. Gregory, the B. & E. Co.'s Engineer and after being tried on the West London Railway was put to work on the Tiverton Branch. The vertical tubular boiler was 3ft. diameter by 6ft. high, the cylinders were 8ins. by 12ins. and drove a dummy crankshaft behind the boiler connected to the 4ft. 6ins. single driving wheels by outside coupling rods. The carriage was mounted on four 3ft. 6in. wheels and provided accommodation for 16 first and 32 second class passengers and for luggage on the roof. The car apparently began work on the B. & E. R. on 1st May, 1849 (as the G.W.R. had provided locomotives to work the line until that date); it worked later on the Clevedon and Weston Branches but did not remain for long in its original condition. In March 1851 the Engineer presented a plan for separating the engine from the carriage. This was adopted and a pair of 3ft. trailing wheels provided.

It is not certain what other alterations were made at this time but an undated drawing shows the locomotive with a horizontal boiler, 8in. by 12in. cylinders, 8ft. 3in. wheelbase, well and back tanks of 160 and 238 gallons capacity, and four wheels quoted as 4ft. but drawn as 3ft.

In August 1855, Messrs. Hutchinson and Ritson, who had a timber-creosoting yard at Bridgwater and were contractors for the Yeovil branch, offered to purchase the *Fairfield* engine for £600 provided that it was converted into a four-wheel coupled locomotive by the Company. This appears to have been done and the engine delivered in June 1856.

WELSH RAILWAYS (1903-11).

Through the kindness of the makers and of some of the members of the Railway Correspondence & Travel Society, it has been possible to illustrate all the types of railmotor introduced on various Welsh railways during the early years of the present century and as it has also been found convenient to tabulate dimensions and building data, extended reference in the text is unnecessary.

ALEXANDRA (NEWPORT & SOUTH WALES) DOCKS AND RAILWAY COMPANY.

Two railmotors were built in 1904/5 by the Glasgow Railway and Engineering Co., Ltd., Govan, and as this firm closed down some years ago it has not been possible to obtain official figures; those reproduced are from information published at the time. The first car is said to have been capable of a speed of 30 miles per hour.

They worked between Caerphilly and Pontypridd, calling at intermediate halts most of which are still in existence and consist of little more than a nameboard at the back of a trackside path. The cars had fixed steps to suit rail-level access.

According to the annual returns, one of the cars was dismantled in 1911 and converted to an ordinary carriage, while the other disappeared during 1917.

BARRY RAILWAY.

The North British Locomotive Co., Ltd. built two cars at their Atlas Works, Glasgow, in 1905, and it will be seen that they closely resembled G.W. cars except that the design of the entrances and the roof ventilators. The coach portions were by R. Y. Pickering of Wishaw.

These cars were used mainly on the Vale of Glamorgan line.

The stock returns show that they were both withdrawn in 1914 and they were converted to trailers the same year. The trailers were Barry No. 177, later G.W. No. 6131, seating 12 first and 56 third class passengers; and No. 178, G.W. No. 4303, seating 12 second (later reclassed third) and 56 third class. In G.W. days they were converted to a gangwayed pair and fitted with the Swindon control system.

CARDIFF RAILWAY.

The Cardiff Railway Co.'s steam railmotors were very similar in appearance to the later G.W. Suburban cars, except that there was a small first class compartment at the leading end, separated from the rest of the coach by a single-door vestibule which had the usual retractable steps.

These cars were obtained for working the passenger service from the Rhymney station at Cardiff to Rhydyfelin, which commenced on 1st March, 1911.

Photograph] [*Hudswell, Clarke & Co., Ltd.*

Rhymney Railway Steam Rail Motor No. 1

L14

Photograph] [*Locomotive Publishing Co.*

Taff Vale Railway Steam Rail Motor No. 1

L15

Photograph] [*Hunslet Engine Co., Ltd.*

Taff Vale Railway Steam Rail Motor No. 3

L16

THE STEAM RAIL MOTORS OF THE CONSTITUENT COMPANIES

(see *pages* **L13** *and* **L14**)

Photograph] [Hunslet Engine Co.

Taff Vale Railway Steam Rail Motor No. 11

L17

Photograph] [Locomotive Publishing Co.

Taff Vale Railway Steam Rail Motor No. 16

L18

Photograph] [R. Stephenson and Hawthorns, Ltd.

Engine Unit of T.V.R. Steam Rail Motor No. 18

L19

THE STEAM RAIL MOTORS OF THE
CONSTITUENT COMPANIES

(see *page* L14)

Photograph] [*Locomotive Publishing Co.*

G.W.R. Auto Train at Trumpers Crossing Halte
Brentford Branch. 0-4-2T No. 1165 L20
(*see page* L16)

Photograph] [*R. H. G. Simpson*

G.W.R. Auto Train at Adderbury,
Chipping Norton-Banbury Line, 1951 L21
0-6-0T No. 5404
(*see page* L16)

Photograph] [*H. C. Casserley*

Hammersmith and City Railway (G.W.R. and Metropolitan Rly. Joint)
Multiple Unit Electric Train at Kensington, Addison Road, August, 1933 L22
(*see page* L25)

Photograph] [*The British Thomson-Houston Co., Ltd.*

G.W.R. Petrol-Electric Rail Car No. 100

L23

Photograph] [*A.E.C. Ltd.*

G.W.R. Diesel Rail Car No. 1

L24

Photograph] [*A.E.C. Ltd.*

G.W.R. Diesel Rail Car No. 2

L25

INTERNAL COMBUSTION RAIL CARS

(*see page* L19)

Two motor and two trailer cars were upplied by the Gloucester Railway Carriage nd Wagon Co., Ltd., three engine bogies (one spare) being built by Messrs. W. Sisson & o., Ltd., Elmbridge Road, Gloucester; elivery was given in February, 1911.

The boilers were constructed to Messrs. isson's orders by Messrs. Abbott & Co., Ltd., Jewark-on-Trent, and were of the vertical aulti-tubular type, the shell being 4ft. 6in. iameter at the foundation ring, tapered lightly for six feet up and the upper portion nlarged to six feet diameter outside.

The spare engine unit was No. 1, the two ailmotor cars and their regular engine units, Jos. 2 and 3, and the two trailers, Nos. 4 and 5.

Only two railmotor cars were mentioned n the annual returns, from which they lisappeared in 1917; the cars were converted o trailers and the engine bogies lay idle for ome years before they were scrapped. The wo railmotor conversions and the two original railers were numbered 142-145 in the G.W.R. railer series in the same order as their original umbers.

PORT TALBOT RAILWAY AND DOCKS COMPANY.

The only steam railmotor acquired as a going concern from an absorbed company was he Port Talbot Railway & Docks Co.'s No. 1, built by Messrs. R. & W. Hawthorn, Leslie & Co., Ltd., Newcastle-on-Tyne in 1906 with coachwork by Messrs. Hurst, Nelson & Co. of Motherwell. The car was lighted by electricity generated by a steam turbine-driven dynamo and the coach portion was very much like the ater G.W. cars in appearance, although there were two separate single passenger doors on each side which divided the car into three almost equal parts, each door having retractable steps. The engine portion was articulated with the car and was cased in to harmonise with the external appearance of the latter; the six-coupled driving unit was unusual.

The Port Talbot Railway was taken over by the G.W.R. on 1st January, 1908, but the railmotor (like the P.T.R. engines) retained its identity and was withdrawn as P.T.R. No. 1 in July, 1920, and sold in the same month, along with the G.W.R. cars already mentioned. It became the Port of London Authority's car No. 3 and worked the same service as the ex-G.W. cars: like them, it ceased work in 1926 and was cut up in 1928. This railcar had been recorded on G.W. Diagram "S" in the railmotor series, from which the dimensions in Table 6 are mainly taken. Some had probably been modified thus by the P.T.R. or at Swindon, as the makers' records give tube heating surface as 590.0 sq. ft., firebox 67.2 sq. ft., and grate area 12.9 sq. ft. The latter also show the weight full as 57½ tons and seating capacity as 70.

An account in *The Locomotive* for May, 1907, gives the seating capacity as 58, so there may have been two modifications; alternatively the makers may have calculated the seating on the basis of three passengers and two passengers on opposite sides of the gangway (see Diesel Railcars, *post*).

RHYMNEY RAILWAY.

The two Rhymney steam railmotors were designed by Mr. C. T. Hurry Riches, the Company's Locomotive Engineer, and comprised small four-coupled locomotives built by Hudswell, Clarke & Co., Ltd., of Leeds, in September, 1907, and coach portions by Cravens Ltd., of Sheffield. They were intended for working, with a six-wheel coach when necessary, on gradients up to 1 in 35.

The cars could be entered from either end by doors three feet wide, approached by moveable steps which could be dropped down by means of a lever. Lighting was by electricity and the Westinghouse automatic quick-acting brake was fitted. Water for the engine was carried in two large cylindrical tanks suspended from the coach underframe, while coal was kept in bunkers on the engine, which were rather like side tanks in appearance. The engines bore number plates 1 and 2 and the coach portions do not seem to have been separately numbered.

Services between Senghenydd, Caerphilly and Machen, involving a daily mileage of seventy were worked by one car, while the other operated between Rhymney and Ystrad Mynach plus a trip from the latter to Merthyr and back, a roster of 102 miles.

Official diagrams show the engine portion of car No. 1 with an extended frame and a pair of trailing wheels added, supporting the front end of the coach portion and making the engine type 0-4-2. There is no record that this alteration was carried out.

One car lasted only three years, the coach portion being fitted with a normal carriage bogie in 1910 and used as an ordinary carriage; the other remained in stock until 1919 when it was similarly treated. The two engine portions were altered to 0-6-0T by extending the frames backwards to accommodate the extra pair of wheels and a rather large combined water tank and coal bunker; the original side bunkers were removed.

The engines will be more fully described when dealing with the locomotive stock of the Rhymney Railway and it will suffice here to record that engine No. 2* was renumbered 120 on conversion in 1910, while No. 1 became 121 in 1919, with further renumbering to 661/2 respectively by the G.W.R. They were both taken out of stock in 1925, No. 661 in February and No. 662 in June.

---

* *The Locomotive* for February, 1911, gives No. 1 as altered to No. 120 but this was in error.

TAFF VALE RAILWAY.

The Taff Vale was the most extensive user of railmotors among the companies absorbed by the G.W.R. and its first car was built at the T.V.'s Cardiff Works in October 1903, the same month that G.W. No. 1 was turned out from Swindon (fig. L15).

The T.V. car, however, only ran its first trials on 29th October and was not put into stock until December 1903.

All the cars had features in common and were designed by the Company's Engineer, Mr. T. Hurry Riches (father of the Rhymney Co.'s Engineer).

Eighteen engine portions and sixteen carriage portions were built in four batches, from four different makers. It will be seen from the illustrations and the dimensions in Table 6 that the first three lots were much the same but that the last lot were much larger. The makers were :—

No. 1.—Taff Vale Railway, Cardiff.

Nos. 2-7.—Avonside Engine Co., Ltd., Bristol. (Coach portions by Bristol Wagon & Carriage Co.)

Nos. 8-13.—Kerr, Stuart & Co., Ltd., Stoke-on-Trent. (Coach portions by Bristol Wagon & Carriage Co.)

Nos. 14-18.—Manning, Wardle & Co., Ltd., Leeds.

No. 1 was the last locomotive to be built by the T.V.R. in its own shops at Cardiff (West Yard).

Incidentally, published statements to the effect that there were further engine portions Nos. 19 and 20 and that No. 19 was Manning Wardle's No. 1681 are incorrect ; the annual returns never showed more than eighteen car engines and Manning Wardle's Nos. 1681 and upwards are otherwise allocated.

The three coach bodies Nos. 14, 15 and 16 were built by the Brush Electrical Engineering Co., Ltd., Loughborough, and had doors to the vestibules in place of the collapsible gates on the earlier cars ; they also had peculiar roof ventilators resembling short sections of clerestories (fig. L18).

It will be noted that the engine numbers determined the vehicle number on the T.V.R., the direct opposite of G.W. practice where the identity of the engine unit was subordinated to the car number.

Since all engines and cars were interchangeable and there were two spare engines, the numbers of the two components did not always coincide. Building dates are :—

Engine No.	*Date Built*	*Engine No.*	*Date Built*
1	10/03	10	4/05
2	7/04	11	5/05
3	8/04	12	5/05
4	9/04	13	5/05
5	10/04	14	4/06
6	11/04	15	5/06
7	12/04	16	5/06
8	3/05	17	5/06
9	4/05	18	6/06

The cars worked over most of the Taf Vale system, including a business service ove the main line between Cardiff (Bute Road) Cardiff (Queen Street), Cathays Halt an Maindy (North Road) Halt. The two halts which still exist, are interesting in having one platform, served only by trains on the outward journey and a passenger from Cathays t Cardiff has to travel via Maindy North Roac Halt.

The services maintained by T.V. railmotors, wholly or in part, are conveniently tabulated as under :—

Aberdare (Mill Street) and Pontypridd.
Aberdare (Mill Street) and Aberdare (Low Level).
Aberdare (Low Level) and Nantmelyn Platform.
Cardiff and Pontypridd.
Cardiff (Bute Road) and Maindy.
Cardiff, Penarth and Cadoxton.
Pontypridd and Nelson.
Pontypridd and Old Ynysybwl.
Pontypridd, Llantrisant, Cowbridge and Aberthaw.
Pontypridd, Porth and Maerdy.
Pontypridd and Treherbert (Blaencwm).

The dimensions quoted in Table 6 for Taff Vale cars Nos. 14-16 (Engine Nos. 14-18) are from the makers. Taff Vale records of about 1908/9 show the boiler diameter as 3′ 10½″, the tube heating surface as 414.21 sq. ft., the firebox heating surface as 50.65 sq. ft., the tank and bunker capacities as 600 gallons and 11 cwt. respectively, and the weight in working order as 45 tons 8 cwt.

The official returns up to December, 1913, showed a total of 16 railcars plus two spare engines ; the engines continued to be shown as totalling 18 until December, 1919, and as " 18, out of use " in December, 1920, but do not appear at all in December, 1921. Nos. 1/2/4, 10/3/4/6-8 were awaiting scrapping at Cardiff in 1919 and four were actually broken up by the G.W.R.

Eight of the coach portions were converted to trailers for use with motor-fitted 4-4-0T and 4-4-2T between 1914 and 1916, and the other eight were altered to ordinary corridor carriages in 1921/22 (see also Page L18).

Table 6. DIMENSIONS OF CONSTITUENT COMPANIES' RAILMOTORS

	A.D.R. No. 1	A.D.R. No. 2 (e)	Barry Nos. 1 & 2	Cardiff Nos. 1—3	P.T.R. No. 1 (a)	Rhymney Nos. 1 & 2	Taff Vale Railway No. 1	Taff Vale Railway Nos. 2—7	Taff Vale Railway Nos. 8—13	Taff Vale Railway Nos. 14—18
Cylinders	9″ × 14″	9″ × 14″	12″ × 16″	12″ × 16″	12″ × 16″	12″ × 16″	9″ × 14″	9″ × 14″	9″ × 14″	10½″ × 14″
Driving Wheels	3′ 0″	3′ 0″	3′ 7½″	4′ 0″	3′ 0″	3′ 6″	2′ 10″	2′ 10″	2′ 10″	3′ 6″
Carrying Wheels	3′ 0″	3′ 0″	3′ 7½″	3′ 7″	3′ 7½″	3′ 7½″	2′ 10″	2′ 10″	2′ 10″	2′ 10″
Boiler, Barrel length	6′ 10⅞″		9′ 1″	9′ 6″	5′ 10″	5′ 11½″	(c)	(c)	(c)	(c)
Boiler, Diameter	3′ 6″		4′ 11″	4′ 6″ & 6′ 0″	4′ 1⅞″	4′ 7⅛″	3′ 4⅛″	3′ 4⅛″	3′ 4⅛″	3′ 9½″
Tubes, Number	274		462	420	210	210	312	304	304	464
Tubes, Diameter	1½″		1⅛″	1¼″	1¾″	1¾″	1¾″	1¾″	1¾″	1⅝″
Heating Surface :										
Tubes, sq. ft.	271		552 (b)	613	591	598.02	299.5	292.3	292.3	413
Firebox, sq. ft.	76		45 (b)	47	69	91.70	39	39	39	63
Grate Area, sq. ft.	7		11.5	11.5	13.1	14.59	8	8.4	8.4	10
Working Pressure, lb. sq. in.	160	160	160	160	170	175	160	160	160	180
Water Capacity (gallons)	350	350	500		600	700	530	550	550	560
Coal Capacity (cwt.)	20	20	15	10	30	20	10	9	9	12
Wheelbase :										
Engine Bogie	8′ 0″	8′ 0″	8′ 0″	8′ 0″	9′ 6″	8′ 6″	8′ 6″	9′ 6″	9′ 6″	10′ 0″
Carriage Bogie	8′ 0″	8′ 0″	8′ 6″	8′ 0″	9′ 0″	8′ 0″	8′ 0″	8′ 0″	8′ 0″	8′ 0″
Total	43′ 10″	48′ 8″	48′ 3″	46′ 0″	63′ 6″	55′ 10″	49′ 3½″	50′ 3½″	50′ 3½″	59′ 8⅞″
Weight Full (d)	40T 0C	43T 0C	50T 15C	43T 8C	49T 10C	55T 2C	33T 2C	35T 14C	35T 14C	44T 11C
Length over Buffers	59′ 0″	64′ 10″	64′ 10″	68′ 11¼″	76′ 9″	71′ 9½″	58′ 9″	58′ 8″	58′ 8″	70′ 3⅞″
Seating Capacity :										
1st Class	—	—	10	16		—	12	12	—	16
3rd Class	52	58	40	48		64	40	36	48	57
Total	52	58	50	64	66	64	52	48	48	73
Date Built	1904	1905	1905	1911	1906	1907	1903	1904	1905	1906
Makers' Numbers	—	—	16466/7	Sissons 971—973	2650	805/6	—	1474-9	898-903	1676-80

NOTES :
- (a) According to G.W. Diagram S. Makers' records differ (see text).
- (b) Makers give total as 598.
- (c) Boiler Barrel in two parts each 2′ $1\frac{3}{16}$″ long.
- (d) Weight in Working Order but excluding Passengers.
- (e) Boiler dimensions probably the same as for No. 1.

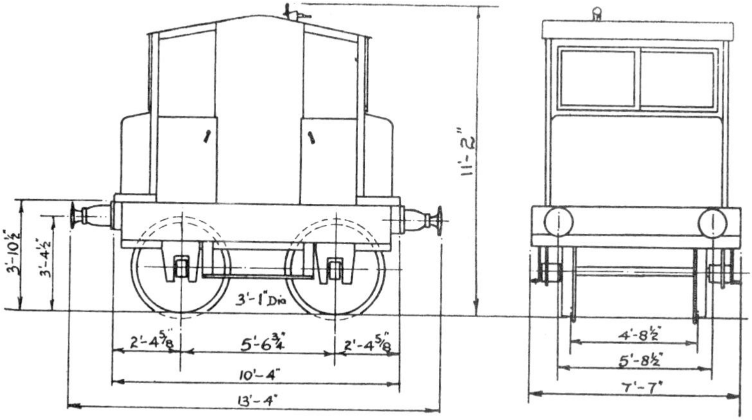

G.W.R. Standard Gauge Simplex Petrol Shunter L26

From a drawing by R. A. Powney

(see page L22)

AUTO TRAIN WORKING

Very soon after the introduction of the first steam railmotor, the G.W.R. started fitting ordinary engines for auto-working. Under this system up to two trailers of the same kind as those used in conjunction with the railmotors could be attached to both the front and the rear of the engine. The driver's cabin of the auto trailers contained a regulator handle which controlled the regulator on the engine through a system of rodding and the driver also had a vacuum brake valve and a remote control for the engine whistle by means of an overhead wire. An electric bell signalling system for communication between driver, fireman and guard was also provided and at a later date a large warning gong, operated by a foot pedal, was fitted outside the control cabin, which suggests that the remote position of the whistle was not always satisfactory although the operating arrangements are still retained. In more recent years some of the auto trailers have been equipped with the Automatic Train control apparatus.

For many years it was the custom to equip the older small engines for auto working, there being several examples of each of the following classes and possibly a couple of the "1901" class 0-6-0ST/0-6-0PT as well :—

"517" Class 0-4-2T.
"Metropolitan" 2-4-0T.
"2021" class, 0-6-0ST/0-6-0PT (4′ 1½″ wheels).
"645," "727," etc. classes, 0-6-0ST/0-6-0PT (4′ 7½″ wheels, single or double frames).

At various times, some of the auto engines were altered or painted to harmonise with the trailers: for example, Nos. 533, 833, 2120/40 were fitted with carriage body casings for a few years. Most, if not all, of them were removed in 1911 but No. 2140 at least retained the carriage contour cab for some years. Later several of the 0-4-2T and 0-6-0T were painted to match the carriage livery of the period, including Nos. 205/22, 518/23/4/6/30/1/3/44/6/61/4, 828/9/31/3/48, 1155-8/60/1/3/5, 1271/5, 1423/5/70/88, 2066, 2101/17/25/33/40.

In the 1920's, a number of the double-framed 0-6-0PT received overall cabs with large square windows in place of the usual round spectacles and were fitted at the same time with screw reversing gear.

In 1930, acceleration of auto services was proposed and 0-6-0PT No. 2080 was from February to April, 1930, experimentally fitted with 5′ 0″ diameter wheels. A further test was made in August, 1930, when No. 2062 of the same class was withdrawn "on paper" and parts of it incorporated in the prototype of a new class of auto engine and numbered 5400. This engine ran exhaustive tests in a number of areas with widely different physical characteristics. The engine was fitted with 5′ 2″ wheels for the trials which took place in September and October, 1930, and with 4′ 7½″ wheels for those held in November and December, 1930.

Based on the experience gained, two standard classes of 0-6-0PT were evolved for auto working, the "5400" class with 5′ 2″ wheels mainly for service in the London area and the "6400" class with 4′ 7½″ wheels for South Wales, Wolverhampton and Plymouth. The old 0-4-2T and 2-4-0T were falling due for

Photograph] [*British Railways*

G.W.R. Diesel Rail Car No. 18

L27

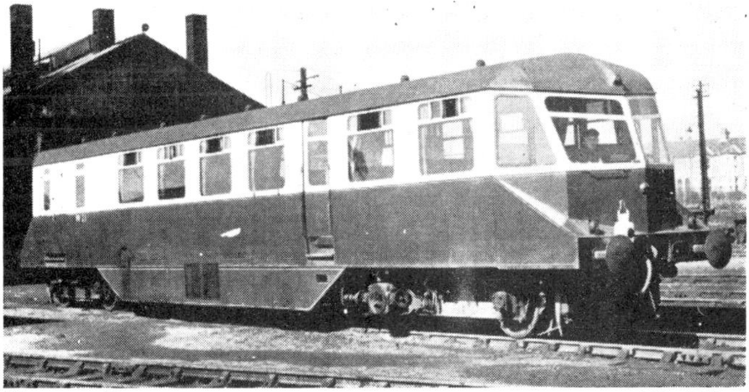

Photograph] [*W. Gilburt*

G.W.R. Diesel Rail Car No. 21

L28

At Weymouth, July 1950

Photograph] [*British Railways*

G.W.R. Twin Diesel Rail Cars Nos. 35 and 36

L29

INTERNAL COMBUSTION RAIL CARS

(*see page* L20)

Photograph] [*P. L. Melvill*

G.W.R. Petrol Shunter No. 15
At Bridgwater Docks, June, 1948 L30

Photograph] [*A. A. G. Delicata*

G.W.R. Petrol Shunter No. 24
At Taunton Engineers' Depot, August 1948 L31

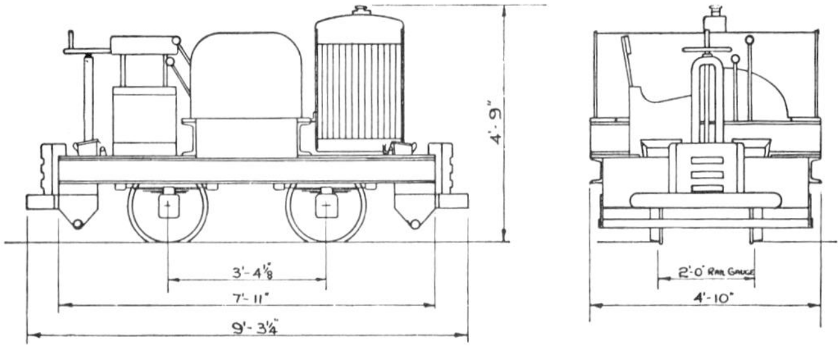

G.W.R. 2ft. gauge Petrol Shunters Nos. 22 and 25
From a Drawing by R. A. Powney L32

PETROL SHUNTING ENGINES
(*see pages* L22 *and* L23)

eplacement about the same time and a nodernised version of the "517" class)-4-2T was brought out to take their place on ›ranches where the new 0-6-0PT would be too ıeavy for the road. These engines were ıumbered 4800-74 (later 1400-74), while the)-6-0PT totalled 65 examples, numbered 5400-24 and 6400-39.

Several of the companies absorbed by the G.W.R. in South Wales had also made use of ıuto engines, sometimes using as trailers the carriage bodies of converted railmotor cars. For example, the Taff Vale fitted up their 4-4-0T, 4-4-2T and some of their 0-6-2T in this way; the system used on the T.V.R. involved a rather unsightly collection of overhead wires and pulleys. The Rhymney Railway used two 2-4-2ST and two 0-6-2ST converted from 2-4-2ST, while the Barry altered at least one of their smallest class of 0-6-0T to 0-4-2T by removing the rear coupling rods and fitted it up for auto working. Some of the Taff Vale 0-6-2T were re-equipped on the standard G.W. system after the grouping.

In September, 1953, the Western Region reorganised the train workings in the Cardiff District, involving a great deal of intensive auto working and to provide additional engine power a number of 45XX and 55XX 2-6-2T were fitted with the apparatus. Their coal and water capacity make long periods of service possible.

AUTO TRAILERS

A few notes on the various kinds of trailer used by the G.W.R. and its constituent companies may not be out of place, especially since a number of them were conversions of steam railmotor cars.

Like the railmotors, the trailers were of two types: with luggage compartments for branch work and without luggage space but with additional end entrance doors for suburban work. When built all cars were glazed at both ends but more recently the trailing ends have been closed in with flush steel sheeting.

The first trailer (No. 1) was built as a companion to railmotor No. 1 and had the same vertical matchboard sides; it was suburban type, 68 seats, 61′ 0″ over buffers. All other newly-built trailers of the earlier batches had panelled sides like the later railmotors and are conveniently summarised as under :—

Nos.	*Type*		*Length over Buffers*	*Seats*
19-24,	Suburban		63′ 6″	64
7, 8, 36-41	Branch			
2, 3, 6,	Suburban		73′ 9″	80
25-28	,,			
48	,,		73′ 9″	96
11-13	Branch		73′ 4″	72
29-34	,,		,,	76
42-47	,,		,,	76
53-70	,,		73′ 11″	76
49-52	,,		73′ 10″	70
71/2, 93-5	Suburban	§	73′ 11″	80
73/4, 96-8	,,	§	,,	76
75-92	Branch	§	,,	70
4, 5	Suburban	§	73′ 6″	76
9	,,	§	,,	80
10	,,	§	,,	80

§*Corridor*

The particulars are taken from the original Diagrams but in some cases the number and arrangement of seats were altered later. The "corridor" notes are from observation and the fitting of gangways to Nos. 3-6 seems to have been carried out after they were built.

Nos. 14-18 and 35 were apparently conversions of clerestory corridor carriages, seating 65 (63 for No. 18, including one First Class compartment) and measuring 55′ 1″ to 57′ 1″ over buffers. Nos. 16, 17, 18 and 35 had lavatory accommodation.

Nos. 99-140 were conversions of steam railmotors and the numbers 141 (Cambrian) and 142-145 (Cardiff) were used for absorbed trailers of general style similar to G.W. cars. The Cardiff cars have 80 seats (143 is still diagrammed as a composite with 16 First and 64 Third Class seats), they are 69′ 0″ long and work in pairs with corridor connections. They had been used as ordinary carriages on the Cardiff Railway and for some time after the grouping, until they were fitted with the G.W. control system. Nos. 146-158 were further railmotor conversions, which were followed by a modern design of branch type flush steel panelled trailer, Nos. 159-170 measuring 63′ 4″ and Nos. 171-80/7-96, 66′ 6″ over buffers. The list is completed by Nos. 181-6, 197-219 which are all converted railmotors. All the foregoing carry their numbers in large figures and are numbered independently of ordinary carriage stock, although there are other types of trailer which are included in the carriage list.

During about 1916-20, the G.W.R. converted a number of bogie suburban coaches with flattened roofs (dating from the 1890's) into trailers. These were known as the "Clifton" type and many were altered from tri-composite coaches (1st, 2nd and 3rd class). They measured 54′ 4″ over buffers and were numbered in the 32XX, 33XX and 71XX series. Most of the absorbed companies had several composite trailers, as well as third class only, and except for Nos. 141-5 already mentioned, the trailers took blanks in the carriage stock list. Except for the Rhymney and Taff Vale, the trailers converted from rail-motor cars have already been described.

The Rhymney Railway did not convert the coach portions of their railmotors to trailers. Car No. 2 became a 64-seat third class carriage, R.R. No. 114 and G.W. No. 1234 while Car No. 1 similarly became R.R. No. 119 and G.W. No. 1236. They were, however, fitted up as trailers on the G.W. system about 1927.

For use with their auto engines, the Rhymney had trailers Nos. 20-2/5/37/42/56, probably conversions of ordinary carriages which became G.W. Nos. 1175, 1000/6, 1176/8, 1078/90 in that order. These do not seem to have been converted to the G.W. system but the Rhymney carriage, G.W. No. 1067, was so fitted.

The Taff Vale trailers are most conveniently tabulated. Nos. 1-16 were of course the former railmotor cars while the remainder were either built as or converted to trailers :—

No.	*Date altered*	*Type*	*Seats 1st*	*Seats 3rd*	*G.W. No.*
1	5/15	D	22	34	6395G
2	/21	T^1	22	34	6396G
3	8/14	D	22	34	6398*
4	/21	T^1	22	34	6399§G
5	/21	T^1	—	60	1308
6	/22	T^2	32 s	32	6475‡
7	/21	T^1	—	60	1310G
8	11/15	D	22	34	6400
9	5/16	D	22	34	6401G
10	8/14	I	—	52	1312
11	5/15	I	—	52	1314G
12	11/15	I	—	52	1315
13	5/16	I	—	52	1317
14	/22	T^2	—	76	4023
15	/22	T^2	—	80	4024
16	/22	T^2	—	76	4025

No.	*Date built or altered*	*Type*	*Seats 1st*	*Seats 3rd*	*G.W. No.*
351	1905b	N	—	42	3871
352	1905b	N	—	42	3872
78	1907b	D	16	70	6421 (Later 2521)G
79	1907b	D	16	70	6422G
353	1907b	D	—	80	2506G
354	1907b	D	—	80	2507G
301	1910a	D	—	50	3855†
302	1909a	D	—	50	3856†
303	1909a	D	—	50	3857†
304	1909a	D	—	50	3858†
331	1910a	I	—	48	2488† (Later 6399)
332	1910a	I	16	32	6430†
80	1912b	I	16	56	6423G
355	1912b	I	—	72	2508G

Notes—

N : Non-auto lightweight Trailers.
D : Driving Trailers.
I : Intermediate (non-driving) Trailers.
† : Not used for auto work after 1920.
G : Later fitted with G.W. type control gear
T^1 : Formed part of 6-coach corridor train 1921, with the addition of former trailers Nos. 8 and 12.
T^2 : Formed part of 4-coach corridor train 1922.
a : Date altered to Trailer.
b : Date built as Trailer.
* : Later altered to 56 3rd class and renumbered 4022.
§ : Later altered to 1st class and re-numbered 8090.
‡ : Altered to 3rd class and renumbered 4022, then to 1st class and renumbered 8091.
s : 32 second class seats.

The last type of trailer to be described was built in 1933-7, with another batch of revised outline in 1938. The length was 60′ 6″ over buffers and the vehicles were of compartment type, similar in appearance to the close-coupled twin sets of the same period except that the guard's compartment was amalgamated with the luggage space to leave room for the driving compartment. These include Nos. 1668-71, 4350/64 and 6818/20.

From trailer No. 159 onwards, electric lighting was standard for new construction, but very few of the earlier trailers have been altered from their original gas lighting which is rather unusual as very few other G.W. vehicles built in the present century retain this form of illumination.

While not strictly part of a Great Western history, mention should also be made of a new series of G.W. design auto trailers built by the Western Region in 1951. These are numbered 220-234 and revert to the saloon type of the earlier cars but with modern sliding ventilators to the windows. These, like most repaints of G.W. trailers, have B.R. standard crimson and cream livery.

In 1952 it was announced that these trailers would carry the names of British birds. In May of that year cars 220/1 received the names *Thrush* and *Wren* respectively, but no further cars were named.

In connection with the extension of auto working in the Cardiff area, in September, 1953, a number of G.W. 3rd class compartment carriages were fitted with through controls only, as non-driving trailers, releasing driving trailers formerly used as " Intermediates " for control purposes. Later an L.M.S. carriage was also fitted as a " G.W. " type intermediate trailer.

From 1955 further control trailers have been provided by conversion of G.W. compartment type brake thirds, which have been renumbered in the 2XX series.

INTERNAL-COMBUSTION RAILCARS

The Petrol-Electric railcar obtained in 1911 was numbered 100 in the railmotor series and was powered by a 40 h.p. Maudslay petrol engine driving a dynamo which supplied current to two electric motors on the axle. Enough petrol could be carried for a run of 250 miles and the maximum speed was nearly 35 m.p.h. The car was designed by the British Thomson-Houston Co., who supplied all the electrical equipment; it was carried on four wheels, and had a total weight of 14 tons 9 cwts. (6 tons 9 cwts. on the leading and 8 tons on the trailing axle). The overall length was 33 feet 3 inches, and the seating capacity is shown officially as 44, although details published in 1912 gave 46 seats. In the latter year it was working on the Windsor branch but it had a short life on the G.W.R., being withdrawn in October, 1919, and sold in the same month to Lever Bros., Ltd., Port Sunlight, Cheshire; the car was in use until about 1923 conveying passengers over their private railway. It was then withdrawn from service and dismantled, but about four years later the body was sold to a member of the staff who converted it into an attractive summer bungalow at Gronant Beach, near Prestatyn, where it still exists.

This vehicle was recorded on Diagram " U " in the railmotor series; the letter " U " was used again later for the Diagram of Diesel Railcar No. 1.

According to information kindly supplied by the Maudslay Motor Co., Ltd., overheating of the detachable valve seatings affected the performance of an otherwise entirely satisfactory petrol engine and this may be the reason why there were no further developments in internal combustion passenger vehicles on the G.W.R. until the advent of the diesel engine. A diesel mechanical railcar built by the Associated Equipment Co., Ltd., of Southall, with bodywork by Park Royal Coachworks Ltd., was put into service on 4th December, 1933, in the Reading area on local journeys. The bodywork is streamlined and the interior finish is typical of omnibuses and coaches of the period, even to the notice that " Smokers are requested to occupy rear seats." The car was taken out of service in the same month for modification to the braking system and was then fitted with automatic train control apparatus, returning to service on 5th February, 1934.

The principal dimensions are tabulated *post* and it should also be mentioned that the power is supplied by a single diesel engine of 121 b.h.p., transmitted to the wheels through a drive enclosed in the axle boxes on the ends of the axle. All the driver's control equipment is duplicated so that the car may be driven from either end and electric lighting is provided throughout including external lights necessary for traffic purposes, although a standard oil tail lamp is carried; these features are common to all the later cars. The sides of the car, as in Nos. 2 to 17, were carried down to within about 12 inches of rail level totally enclosing the underframe although several sections were removable for access. Most, if not all, of the cars are now running with parts of this valance removed from the driving side of each bogie (sometimes from both sides of each bogie) and from the engine.

Car No. 1 has its seats arranged back to back on each side of an off-centre gangway, accommodating three passengers on one side and two passengers on the other side of the gangway. There are two passenger compartments, one on each side of a centre entrance vestibule and a driver's compartment at each end, one of which is made larger to accommodate luggage as well; all the doors open outwards. Each passenger compartment has a heater fed from the radiator.

The next three cars were more ambitious as they were designed for express services, with a maximum speed of 75 to 80 miles per hour, compared with No. 1's top speed of 60 miles per hour. Many of the fittings of the first car are retained but two engines of 121 b.h.p. are employed, each driving a separate bogie. The entrance vestibule opens as before into two compartments but the gangway is central with two seats on each side, similar to the modern type of restaurant car, including tables between the seats. Five groups of seats for eight passengers are provided and the gangway from the smaller compartment leads past two flanking lavatory cubicles to the buffet section, which contains four seats; the remainder of the car is given over to the luggage compartment and the two driver's cabins.

These cars were equipped for Automatic Train Control when built and this feature was also fitted to all subsequent cars when new. The vacuum brake is standard on all vehicles but up to car No. 17 it is self-contained and these seventeen cars are fitted with buffers of a special design and concealed drawgear making it possible for a car to be towed in case of a breakdown.

The three buffet railcars were used on a Cardiff-Birmingham express service from 1934 until about 1940 and on Swansea-Cheltenham and Newport-Swansea express services after the war. They were either laid up or used on local trips during the war and were replaced on the Swansea-Cheltenham service by cars of the 10-12 series which have a higher seating capacity. These in turn gave way to steam trains upon reorganisation of the services. No. 2 was damaged in a collision at Newport on 7th December, 1953, and withdrawn on 8th February, 1954.

Commencing with No. 5, the car bodies were built by the Gloucester Railway Carriage and Wagon Co., Ltd., although the same streamlined outline was retained. Cars Nos. 5 to 16 (inclusive) are generally similar to No. 1 except that they have twin engines each of 121 b.h.p.; Nos. 6 and 7 have single gear-boxes (like Nos. 1 to 4) but the remainder have two. All these cars have sliding doors and Nos. 10, 11 and 12 have a single lavatory compartment opening off the centre vestibule. A photograph of No. 9 shows it with a warning bell of the type used on auto trailers, in addition to the usual twin horns. This must have been experimental as the feature does not seem to apply to any other cars. No. 9 was almost completely destroyed by fire at Heyford on 24th July, 1945, although not officially withdrawn until 11th May, 1946.

This batch has seen service in a number of areas of the G.W. system and the cars have been transferred in many cases from their original sheds but their more interesting duties have included Oxford to Hereford, Bristol to Cardiff and Bristol to Weymouth express business services: the Bristol-Cardiff service has been discontinued while in the other cases the traffic has outgrown the cars.

The next car, No. 17, was adapted for parcels traffic and used in the London area. The object was to cater for heavy traffic previously conveyed by passenger trains, thus enabling the latter to be accelerated. The design is identical with the other cars except that there is no passenger accommodation, three sets of glazed sliding doors on each side replacing the normal arrangement.

The eighteenth railcar incorporated several new features as it was designed to haul a tail load of up to 60 tons of passenger or goods vehicles. Standard passenger-type buffers and drawgear and hose couplings for the vacuum brake and steam heating are fitted while a Bowden wire cable and special electrical connections enable the throttle and all other controls respectively to be operated from the driving compartment of a trailer. The car itself can also be driven from either end and all the controls except the throttle are electro-pneumatically operated. The arrangement of the drive is the same as in the previous cars but the bogies are of a stronger design in view of the heavier duties and the valance is confined to the under-part of the car between the bogies. Seating accommodation is limited to 49 in the normal five-a-side style of most of the earlier cars and the luggage compartment is increased in proportion; extra seating capacity is of course given by a trailer when required. Car No. 18 was built by the same firms as Nos. 5 to 17 and the two engines were again of 121 b.h.p. each.

A further batch of twenty cars was authorised in February, 1938, which were also to be fitted with the standard drawgear, etc. of the experimental car No. 18. The engines were again built by A.E.C. but the bodies and undercarriages were made at Swindon and the streamlined outline of the earlier cars was abandoned in order to give more usable floor space. The brake gear was also modified, with the brake-blocks acting on the wheel tyres (as in usual railway practice) instead of on the brake drums which were fitted to the first eighteen cars.

Fifteen of the cars are intended for branch and subsidiary services and are geared down to give a maximum speed of 40 m.p.h. in order to provide sufficient tractive effort for hauling a tail load of up to 60 tons. Nos. 19 and 20 are however fitted with special dual-range gear-boxes to permit an alternative maximum speed of 60 m.p.h. where traffic conditions and ruling gradients allow. The general layout of the cars is as before except that the gangway is central with groups of four seats on either side. These and the other five cars of the new batch have a hand throttle lever on both sides of the cab, which can be worked by the driver when leaning out of the window. A single upward or downward movement of either lever now gives complete control of the car from starting to braking, thus making it easier for the driver to perform shunting operations or to pick up the train staff on single lines.

Car No. 34 is for parcels traffic and apart from the later underframe and body contour is generally similar to No. 17. It is used in the London area (fig. L1).

The last four cars were designed as two pairs of twin units with driving compartments at the extreme ends only to replace the earlier buffet cars on the Birmingham-Cardiff service. Remote control enables both cars of a set to be driven from either end by one man and this principle has been further extended by introducing a standard carriage in between the two diesel units to form a three-car train. The top speed is 70 m.p.h. and Cars No. 36 and 38 have a buffet counter, while Nos. 35 and 37 have lavatory accommodation, a standard gangway being provided for access from one vehicle to the other. The two-compartment arrangement is again followed with access from a centre vestibule but the smaller compartment in the buffet car seats only 12 passengers, with 32 in the other. The other car in each set has 28 and 32 seats respectively and all four cars have a luggage compartment.

Despite the addition of the third vehicle the Birmingham-Cardiff traffic developed to such an extent that the Diesel sets were soon only operating mid-week and they were eventually replaced entirely by steam trains (which

Diesel Electric Shunter No. 2 (now B.R. 15100)
At Swindon, 1936

L33

English Electric Diesel Electric Shunter No. 15101

L34

Brush Diesel Electric Shunter No. 15107

L35

From British Railways Photographs

DIESEL ELECTRIC SHUNTING ENGINES

(see *page* L24)

Photograph] [*K. R. Prentice*

G.W.R. Fowler Diesel Mechanical Shunter No. 1

(*see page L23*)

L36

Photograph] [*British Railways*

Brown Boveri Gas Turbine Locomotive No. 18000

(*see page L24*)

L37

Photograph] [*British Railways*

Metropolitan-Vickers Gas Turbine Locomotive No. 18100

(*see page L25*)

L38

make up to no less than fifteen vehicles on exceptional occasions), whereupon one set worked between Bristol and Weymouth and the other between Reading and Newbury. When one or other of the twin cars has not been available, one of the other trailer-hauling cars has been used to complete the set although the maximum speed was presumably limited to that of the slower partner.

Car No. 37 was damaged by fire in 1947 and was stored until it was taken out of stock on 23rd September, 1949. Its place in the twin set was filled by No. 22 from 19th April, 1948, but No. 33 was rebuilt 24th March, 1954, with a single driving end and vestibule connections for permanent use in this way.

Before setting out dimensions and dates, it should be mentioned that Cars Nos. 6 and 19 were on loan to the L.N.E.R. for a while during 1944, when they were used on local services in the Newcastle area and that all cars are having their numbers prefixed with the letter W when repainted by British Railways in standard crimson and cream livery or crimson only in the case of Cars 17 and 34.

The coupling shafts on the bogies of the later cars have most, if not all, been removed so that the drive is on one axle to each bogie only.

The principal dimensions and details are conveniently tabulated as under ; the official Individual Dates to Stock are probably those of acceptance, while the cars were presumably added in the annual returns in the years in which they commenced work on the G.W.R. Nos. 1 to 18 have A.E.C.-Ricardo engines while those on Nos. 19 to 38 are of the A.E.C. direct injection type.

Allocation of the cars in June, 1947, was : Southall 2, Reading 5, Oxford 2, Bristol 4, Weymouth 1, Leamington 2, Stourbridge 3, Gloucester 1, Worcester 7, Newport 2, Llantrisant 2, Pontypool 1, Carmarthen 2, Landore 3.

Staff difficulties in the London area resulted in a number of transfers into that District, while Llantrisant and Pontypool lost their railcars due to branch line closures. The distribution in October, 1955, was : Southall 5, Reading 4, Oxford 4, Bristol 5, Weymouth 1, Leamington 3, Stourbridge 3, Worcester 6, Gloucester 1, Newport 2, Landore 1.

TABLE 7

SUMMARY OF DIESEL RAILCARS

Car No.	*Carriage Lot No.*	*Individual Stock dates*	*Added to Stock in*	*Weight empty* T	C	*Seating Capacity*	*Railcar Diagram*
1	1516	5/2/34	1934	24	0	69	U
2	1522	9/7/34	,,	26	4	44	V
3	,,	17/7/34	,,	,,		,,	,,
4	,,	26/9/34	,,	,,		,,	,,
5	,,	22/7/35	1935	25	6	70	W
6	,,	30/8/35	,,	,,		,,	,,
7	,,	8/7/35	,,	,,		,,	,,
8	1547	5/3/36	1936	29	10	,,	,,
9	1546	3/2/36	,,	,,		,,	,,
10	1547	17/2/36	,,	29	18	63	X
11	,,	,,	,,	,,		,,	,,
12	,,	11/2/36	,,	,,		,,	,,
13	1546	16/3/36	,,	29	10	70	W
14	,,	23/3/36	,,	,,		,,	,,
15	,,	6/4/36	,,	,,		,,	,,
16	,,	17/4/36	,,	,,		,,	,,
17	1547	27/4/36	,,	28	17	*	Y
18	1564	4/4/37	1937	33	12	49	Z
19	1635	11/7/40	1940	35	13	48	A[1]
20	,,	4/6/40	,,	,,		,,	,,
21	,,	11/7/40	,,	,,		,,	,,
22	,,	18/9/40	,,	,,		,,	,,
23	,,	,,	,,	,,		,,	,,
24	,,	,,	,,	,,		,,	,,
25	,,	,,	,,	,,		,,	,,
26	,,	,,	,,	,,		,,	,,
27	,,	6/11/40	,,	,,		,,	,,
28	,,	16/12/40	,,	,,		,,	,,
29	,,	6/1/41	,,	,,		,,	,,
30	,,	13/1/41	,,	,,		,,	,,
31	,,	3/2/41	,,	,,		,,	,,
32	,,	,,	,,	,,		,,	,,

TABLE 7 (*continued*)

Car No.	Carriage Lot No.	Individual Stock dates	Added to Stock in	Weight empty T C	Seating Capacity	Railcar Diagram
33	1635	3/3/41	1941	35 13	48	A[1]
34	1636	15/9/41	,,	34 18	*	A[2]
35	1637	16/11/41	,,	36 14	60	A[4]
36	,,	18/11/41	,,	37 12	44	A[3]
37	,,	28/2/42	,,	36 14	60	A[4]
38	,,	,,	,,	37 12	44	A[3]

*Nos. 17 and 34 convey parcels only, capacity 10 tons.

Car No.	1	2-17	18	19-38
Cylinders (6), bore (mm.)	115	115	115	120
,, ,, stroke ,,	142	142	142	142
Brake horse-power . .	121	242	242	210
Wheels, diameter . .	3′ 1″	3′ 1″	3′ 2″	3′ 2″
Bogie, wheelbase . .	7′ 0″	7′ 0″	8′ 6″	8′ 6″
,, centres	40′ 0″	40′ 0″	43′ 6″	43′ 6″
Length over buffers . .	63′ 7″	63′ 7″	65′ 8″	65′ 8″

INTERNAL-COMBUSTION SHUNTING LOCOMOTIVES

Petrol Shunting Engines, Nos. 15, 22-27

The G.W.R. is believed to have purchased some Drewry petrol-engined inspection trolleys before 1910 and a " Simplex " locomotive before 1921 but no details are available and numbers do not seem to have been allotted in the steam engine series. The first recorded example of such an engine numbered in the locomotive stock occurs in 1923, when a four-wheeled " Simplex " petrol shunter was obtained from the Motor Rail and Tram Co., Ltd., of Simplex Works, Bedford. It was ordered by the Locomotive Department as G.W. No. 15 under Lot 229 but this Lot was later cancelled and the engine handed to the Goods Department, who appear to have received it in April, 1923. The engine was used at Wednesbury until 1932, and thereafter at Bridgwater and Taunton until 1939 but was then laid aside and transferred to " Spare Machinery Account " in September of that year. More recently it was at Taunton again but arrived derelict at Swindon 15/2/51 and orders were given for cutting up on 1/3/51 (fig. L30).

Another engine, ordered on 17th March, 1923, was built under Simplex Works No. 2215 but the order was cancelled before delivery and the engine stored by the makers until 1925 when it was sold to the Dunstable Portland Cement Co. Ltd. under Works No. 3786. It has since been seen at the Washington Chemical Co.'s premises, but bearing Works No. 2215. The foregoing is noted to prevent confusion but the engine in question can hardly be regarded as a " Locomotive of the G.W.R."

Three similar shunters were built by the same firm for the Docks, Engineering and Signal Departments respectively and put to stock in 1926, no Lot Number being issued in this case. A fifth engine was obtained under Lot 250 for the Provender Stores at Didcot in 1927. In December, 1928, all five engines were shown as Service Stock and the four last described received the numbers 23, 24, 27 and 26 respectively in November, 1932, the numbers being painted on. No. 15 had always had a G.W. standard number plate but with the small letters " G.W.R." above the numbers on the style adopted for engines taken over from absorbed companies in 1922-24.

Nos. 15 and 23 are recorded on Diagram "A" in the G.W. internal combustion series but Nos. 24, 26 and 27 do not appear to have a diagram index letter.

No. 23 has worked mostly at Bridgwater Docks while Nos. 24, 26 and 27 have been almost continuously at Taunton Engineers' Dept., Didcot Provender Stores and Reading Signal Works respectively.

The five engines are of generally similar design except that No. 15 had completely open sides to the cab while the others have partly sheeted sides with two cab doors on each side of the body. Standard buffers and three-link couplings are fitted and there are large sliding look-out windows at each end of the cab, which vary in detail between the engines built under the three separate orders. No. 23 has acquired a warning bell mounted on the cab roof, of the type used on steam railmotors and auto trailers.

These petrol shunters weigh 8 tons and have 4-cylinder petrol engines developing 40 B.H.P. at 1,000 R.P.M., with a chain drive on to both axles through a two-speed gearbox. The low gear provides a tractive effort of 3,400 lb. at a speed of 3 m.p.h., the corresponding figures for high gear being 1,540 lb.

t 7.2 m.p.h.; the same speeds apply when he shunter is travelling in either direction. The general arrangement and principal dimensions are shown on the accompanying line drawing (fig. L26).

Building details are as follows :—

G.W. No.	Maker's No.	Date built*	Date to Stock	Consigned ex-Works to
15	2138	3/23	1923(?)	Stafford Road Depot, Wolverhampton
23	3731	5/25	1926	Loco. Dept., Bridgwater
24	3821	1/26	1926	Engineer's Dept., Taunton
26	4178	4/27	1927	Provender Stores, Didcot
27	3820	1/26	1926	Signal Works, Reading

*According to Maker's records

——::——

Two 2ft. 0in. gauge "Simplex" shunters were built 14/1/30 (Makers' No. 5031) and 3/5/30 (Makers' No. 5074) and numbered 22 and 25 respectively as Service Rolling Stock in 1932. They were despatched ex-works to the Engineer's Department at Banbury and St. Austell and were evidently intended for use on Jubilee track on construction jobs. No. 22 was later reported at Westbury and No. 25 at Perranwell. For some odd reason they carried numbers in the steam locomotive stock, although they were almost immediately taken out of the Service Stock list and transferred to "Machinery and Plant Stock." They never appeared in the Service Stock Annual Totals. No photographs are available but the line drawing shows the general appearance and gives leading dimensions. In January, 1951, both were reported at Hayes Sleeper Creosoting Depot where they bore numbers P.W.M. 1780 and 1779 respectively.

The two-cylinder petrol engine develops 20 B.H.P. at 1,000 R.P.M. giving a tractive effort in low gear of 1,250 lb. at 3.5 m.p.h., the figures for high gear being 956 lb. at 7.0 m.p.h. The weight is 2½ tons (fig. L32).

The Engineer's Department also possesses two 2ft. 0in. gauge Diesel shunters of somewhat similar design, Simplex No. 7139/1936 and 7177/1937. The latter was also at Hayes in January, 1951, numbered P.W.M. 1650 but left there in company with P.W.M. Nos. 1779/80 in the following month and No. 1780 was in store at Theale in June, 1953. Their present locations are not known.

Diesel Mechanical Shunter No. 1

A six-cylinder four-wheeled Diesel-Mechanical shunting engine with an auxiliary petrol engine for starting was purchased in April, 1933, from John Fowler and Co. (Leeds) Ltd. under Lot 289 and put into Departmental Stock as No. 1 in the locomotive stock series. This engine had been designed by the makers (their No. 19451 of 1933) and was used by them as a demonstration locomotive for a short time prior to being sold to the G.W.R. It was one of the few G.W. engines to have Westinghouse brakes and worked in the yards at Swindon. In September, 1939, it was transferred to the Spare Machinery Account and sold in March, 1940, to Geo. Cohen, Sons and Co., Ltd., Stanningley, Leeds, who resold on 12th October, 1940, to the Ministry of Supply.

The official I.C.E. Diagram B records the following dimensions :—

B.H.P. of engine ..	70
Cylinders, bore and stroke	120 mm. × 180 mm.
Wheels, coupled ..	3′ 0″ diameter.
Speeds, forward and reverse	3, 6, 10.2 and 15 m.p.h.
Tractive effort ..	7,000 lb. at 3 m.p.h. 3,500 lb. at 6 ,, 2,060 lb. at 10.2 ,, 1,400 lb. at 15 ,,
Fuel Oil Tank Capacity	46 gallons.
Wheelbase	5′ 6″
Length over buffers	22′ 1¼″
Weight full	18 tons 11 cwt.
Minimum curve ..	80 feet radius.

Diesel Electric Shunting Engines

The first Great Western Railway diesel-electric shunter was purchased under Lot 302 and put to stock in April, 1936, having been built the previous year by Messrs. R. and W. Hawthorn, Leslie and Co., Ltd., under their Works Number 3853. This firm made the frames and mechanical equipment while the electrical gear was supplied by the English Electric Company Ltd. The engine was numbered 2 in G.W. stock and was used at Acton for many years although it was sent to Swansea during the war where it shunted the oil wharves in company with L.M.S. No. 7069 (later replaced by No. 7076). It returned to Acton later and became British Railways No. 15100 in February, 1948; more recently it has been moved to Bristol. Incidentally this was the only G.W. engine required to be renumbered under the British Railways scheme.

The primary source of power is a six-cylinder vertical type diesel engine, fed by fuel pumps and injectors and water-cooled. The water circulation is driven by a centrifugal pump around the engine and radiator jackets. The radiator fan is controlled by thermostats inserted in both the oil and water systems. Lubrication is by an automatic pressure system which includes a safety device whereby the engine is shut down if the oil pressure falls.

The electrical equipment comprises a 230 KW. main generator coupled direct to the engine and feeding current to two motors driving on to the end axles. Coupling rods transmit the drive to the centre axle and there is an arrangement for running on one motor only in the event of the other failing.

The engine has a maximum speed of 19 m.p.h. on the level, when it can haul 500 tons but the load can be increased to 1,000 tons at half this speed. On a 1 in 100 grade the maximum load is 750 tons at 2.7 m.p.h.

The cab is totally enclosed, with large observation windows, and the principal controls can be operated from either side of the cab. In Great Western days the body casing was painted green and lined out but the present colour scheme is the British Railways standard unlined black for shunting locomotives (fig. L33).

Under the 1946 renumbering scheme, No. 2 was to have become No. 500, while Lot 363 provided for one diesel shunter to be numbered 501 with the power units supplied by the Brush Electrical Engineering Company and Lot 364 covered six similar machines (Nos. 502-7) but with English Electric power units. Everything except the power units was to be built at Swindon.

Before these orders were completed, th British Railways numbering scheme came int force and Lot 364 came out as Nos. 15101- while the engine originally intended to b No. 501 appeared as No. 15107. The engine were put into stock as follows :—

1948 : 15101/2 (April), 15103/4 (May) 15105/6 (July).
1949 : 15107 (November).

The whole of Lot 364 were painted G.W green (unlined) and lettered "Britis Railways" in G.W.-style block capitals an all were allocated to Old Oak Common, bein employed in that area. No. 15107 came ou in unlined black paint with the Lion an Wheel emblem and was sent to Bristol. Al seven engines have G.W. style number plates

The design of these newer locomotive generally followed that of No. 2 and th English Electric power units have six cylinder like the original example, but the Brush uni has four cylinders. A standard arrangemen of framing and bodywork was adopted fo Nos. 15101-7 and No. 15107 is indistinguishabl externally from the other six (figs. L34 and L35).

Leading dimensions and details are as follows :—

All the engines stand on a wheelbase of 11′ 6″, equally divided ; No. 15100 measures 29′ 0½″ over buffers while the others are 1″ longer.

	15100	15101-6	15107
B.H.P. of Engine	350	350	360 (2 stroke)
Cylinders, bore and stroke ..	10″ × 12″	10″ × 12″	8½″ × 13″
Speed of Engine	675 R.P.M.	680 R.P.M.	600 R.P.M.
Wheels, diameter	4′ 0½″	4′ 0½″	4′ 0½″
Main Generator	230 KW	190 KW	190 KW
Auxiliary Generator	11 KW	4 KW	9 KW
H.P. of Traction Motors ..	170 each	135 each	116 each
No. ,, ,, ,, ..	2	2	2
Main Fuel Oil Tank Capacity ..	400 gallons	584 gallons	584 gallons
Auxiliary Fuel Oil Tank Capacity	100 ,,	75 ,,	75 ,,
Starting Tractive Effort ..	30,240 lb.	38,500 lb.	35,000 lb.
Weight, Full	51 tons 11 cwt.	46 tons 9 cwt.	46 tons 2 cwt.
,, Empty	48 tons 0 cwt.	43 tons 0 cwt.	42 tons 18 cwt.
Engine Diagram (I.C.E. series) ..	C	D	E

GAS TURBINE LOCOMOTIVES

In 1946, the G.W.R. ordered a gas turbine locomotive from the Swiss firm of Brown-Boveri, based upon that firm's original engine which had been in service in Switzerland since September, 1941. It is the third example in the world, an American-built locomotive for the Union Pacific Railroad having been completed a few months earlier than the G.W. locomotive which appears to have started running trials in Switzerland during the summer of 1949 and arrived at Harwich on 3/2/50, whence it was despatched dead to Swindon. It was not put to stock until 9/5/50, but prior to this date it had worked a number of trial trips and, as might have been expected with a new design, had some teething troubles, including damage to the turbine. By 20th May, 1950, it had worked a number of passenger turns between Swindon and Paddington and a few days later entered into regular express duty between Paddington and Plymouth and Paddington and Bristol and has since been chiefly employed on the latter services. No. 18000 is restricted to the same routes as apply to "King" class 4-6-0's.

Lot 372 was issued to cover this order to Brown-Boveri, who sublet the mechanical parts to the Swiss Locomotive Company. Delivery was not made until after the formation of British Railways and the number 18000 was allotted in the B.R. internal-combustion locomotive series.

Details are :—

No.	Makers	Maker's No.	Date to Stock
18000	Brown Boveri / Swiss Locomotive Co.	4559/1949 / 3977	5/1950

A gas turbine works by inhaling large quantities of air into a compressor and heating this air by passing it through a combustion chamber in which fuel oil is burned, the air then emerging from the combustion chamber at high velocity to drive a turbine. In the case of No. 18000 the exhaust from the turbine passes through a heat exchanger where some of the waste heat is transferred to the air before it enters the combustion chamber, thus saving fuel. The turbine drives an electric generator through gearing and the current produced drives four traction motors, two in each bogie.

Heavy fuel oil (sufficient for 250 miles run), lubricating oil and water are carried partially in the roof and partially in the middle portion of the main frame and (together with sand and other stores) weigh 17,000 lb.

The centre portion of the locomotive is occupied by the power unit, which is mounted on an auxiliary frame and comprises the compressor, heat exchanger, combustion chamber, gas turbine and geared generator with the air intakes located near the centre of the vehicle. An auxiliary diesel-generator set is placed at one end of the power unit in company with the train heating boiler and one of the traction motor fans ; at the other end are situated the exhauster-compressor sets for the brakes, which are duplicated for mechanical and electrical operation when running and standing respectively, together with the other fan. The auxiliary generator provides electric current for all the auxiliaries and is also powerful enough to supply current to two of the four traction motors for working the locomotive when running from the engine shed to the station to pick up its train. The steam for train heating is supplied by a steam boiler automatically fired with oil.

The locomotive can be driven from either end from totally enclosed cabs fitted for remote control of the gas turbine power unit, the diesel generator set and the auxiliary apparatus.

The principal dimensions are : Driving wheels, 4′ 0½″ diameter, carrying wheels 3′ 2″ diameter ; length over buffers 63′ 0″, height above rails 13′ 4″ ; continuous rating of gas turbine unit 2,500 h.p. ; total weight in working order 258,000 lb. with 172,000 lb. available for adhesion on the four driving axles. The locomotive is capable of a maximum speed of 90 m.p.h., when the tractive effort at wheels is 8,400 lb., increasing to 12,400 lb. at 64 m.p.h. The maximum tractive effort is developed at starting and is 31,500 lb. at speeds of up to about 21 m.p.h.

A second gas turbine locomotive was ordered at about the same time from Metropolitan-Vickers Electrical Company, Ltd.. and later covered by Lot No. 388. The makers are the designers and the financial and technical responsibility for the project is shared by them and British Railways.

The locomotive is B.R. No. 18100 and was delivered to Swindon on 16th December, 1951, and taken into stock in April, 1952.

No. 18100 is somewhat similar to No. 18000 in the general arrangement of the plant and control equipment but all three axles on both bogies have driving motors whereas the centre axle of each bogie is idle in the case of No. 18000.

Dimensions so far made public are : Total length 66′ 9¼″ ; height above rails 12′ 10″ ; continuous rating of gas turbine unit at rails, 2,450 H.P. (3,000 H.P. at turbine) ; total weight in working order 129 tons 10 cwt. Like No. 18000, the maximum speed is 90 m.p.h. ; the maximum tractive effort is 60,000 lb. which makes it the most powerful passenger locomotive on British Railways.

The wheel notation of 18000 is A-1-A+A-1-A and 18100 Co+Co. In the Swindon I.C.E. Diagram series, 18000 is diagram F and 18100 diagram G.

THE ELECTRIC RAIL MOTOR VEHICLES

As has been stated the only electric rail motor vehicles comprised a half share in forty electric motor coaches for multiple unit working on the Hammersmith and City Railway which was jointly owned by the G.W.R. and the Metropolitan Railway. The vehicles were of Metropolitan Railway design of the period and numbered in that company's series, Nos. 2208-47. They were maintained by the Metropolitan Railway and as their G.W.R. existence was little more than a book entry it is not considered appropriate to refer in detail to them as G.W.R. locomotives. The transfer of the Hammersmith and City Railway to the London Passenger Transport Board did not affect the position so far as the G.W.R. was concerned but when the older H. and C. coaches were withdrawn in favour of the modern L.P.T.B. metadyne stock in 1938, the G.W.R. as from that year ceased to record any electric rail motor vehicles (fig. L22).

Engraved brass plated fitted to G.W.R. Steam Rail Motors (*see page L4*).

Photographed from a plate in W.E. Hayward's collection.

L39